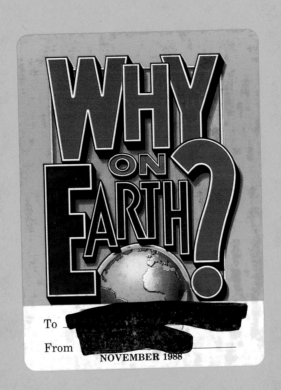

To

From

NOVEMBER 1988

WHY, ON EARTH?

BOOKS FOR WORLD EXPLORERS
NATIONAL GEOGRAPHIC SOCIETY

Contents

TITLE PAGE: Passengers riding a hot-air balloon get a bird's-eye view of the world around them. Why do hot-air balloons rise? You'll find the answer on page 54.

GARY GREENE

2

Copyright © 1988
National Geographic Society
Library of Congress CIP data: 95

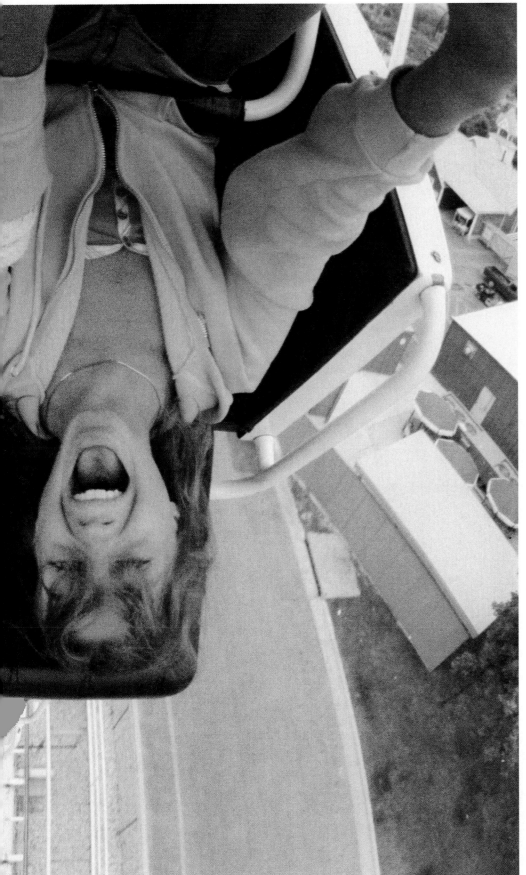

Did someone print this photograph upside down? Are these people really topsy-turvy? If so, why don't they tumble out of their seats?

These are good questions, and they have good answers. The photograph is not upside down. The happily screaming people are. Heads down, bottoms up, they are riding through a loop-the-loop, a circle in a roller coaster track.

All objects tend to travel in a straight line. This tendency, called inertia (in-ER-shuh), works to keep the cars and riders moving straight ahead. But the roller coaster track isn't straight. It curves. The curve of the track and the pull of gravity exert centripetal (sen-TRIP-uh-tul) force. This force opposes inertia and pushes the cars and riders in a circular path.

As the cars race, upside down, across the top of the loop-the-loop, why don't the riders fall out? The speed of the cars and the curve of the track work to overcome gravity and to press the passengers into their seats.

In this book you'll discover a lot of good questions. Many were asked by curious readers like you. Why does skin wrinkle? Why don't ducks sink? Why does the earth move around the sun? How can a sponge hold water if it's full of holes? Do skaters and dancers get dizzy when they spin? No matter how difficult these questions may seem at first, you'll find that each has a logical answer. And why not? After all, doesn't every good question deserve a good answer?■

GROWING CONCERNS

Why did some dinosaurs grow to giant size?

"Hello up there!" Visitors to a museum in Chicago, Illinois, gaze at the skeleton of *Albertosaurus libratus*. This dinosaur probably weighed about three tons—as much as an elephant.

Some dinosaurs were as small as chickens. But many were huge, much bigger than *Albertosaurus libratus*. A dinosaur called "Supersaurus" may have been one of the largest. A single Supersaurus probably weighed as much as 15 elephants!

Dinosaurs roamed the earth for about 140 million years. Then, about 65 million years ago, they became extinct. Today, scientists study dinosaur bones and footprints that have been preserved as fossils to learn more about the creatures.

Scientists don't know exactly why some dinosaurs grew so huge, but they do have several possible explanations. Perhaps certain big dinosaurs never stopped growing from the time they were born until the time they died. Perhaps some dinosaurs grew very rapidly. The duckbill dinosaur, researchers believe, grew from less than a pound at birth to about 6,000 pounds before reaching age ten. Giant dinosaurs may have found food, conserved body heat, and defended themselves better than smaller creatures did.

Researchers have found evidence that during the time of the dinosaurs the earth's climate was different from that of today. Some scientists think a dinosaur's large size shows that the creature adapted to life in a dry climate. Others think that a mild and constant climate made it possible for some dinosaurs to grow into the largest animals ever to walk the earth.■

© LAWRENCE MIGDALE, SCIENCE SOURCE/PHOTO RESEARCHERS

How can I tell a horse's age by looking at its teeth?

If you look for information about a horse's age, you can get it straight from the horse's mouth. A horse has at least 36 teeth. The 12 front teeth—6 uppers and 6 lowers—contain many clues to the animal's age.

By the time the front teeth have grown in as baby teeth, a horse has reached its first birthday. A $2\frac{1}{2}$-year-old horse often has four adult front teeth. A horse has eight such teeth by age four, and twelve by age five.

As the animal ages, its teeth wear down. Two bottom front teeth wear down by age six, four by age seven, and all six by age eight. After that, experts study the changing shape, slant, and markings of a horse's other teeth to figure out how old the animal is.■

ANIMALS ANIMALS/REED/WILLIAMS

Why do I grow in spurts?

"Stand up straight." A father measures his daughter's height to see how fast she is growing.

Like this girl, you have already experienced your first growth spurt—a time of very fast growth. It occurred between birth and age two. When it was over, you probably had reached half your adult height. After age two, your rate of growth slowed down.

Around age 11 to 13, puberty begins. A child begins to develop an adult body. This is another growth spurt. During puberty, you may shoot up four to six inches in a year!

The pituitary (pih-TOO-uh-tair-ee) gland at the base of the brain produces chemicals called hormones. They control how fast you grow. When your body produces more hormones, you grow more quickly. After puberty ends, the pituitary gland produces fewer hormones. At that time, usually in the middle to late teenage years, you stop growing. You've reached your full adult height.■

SLEEPYHEADS

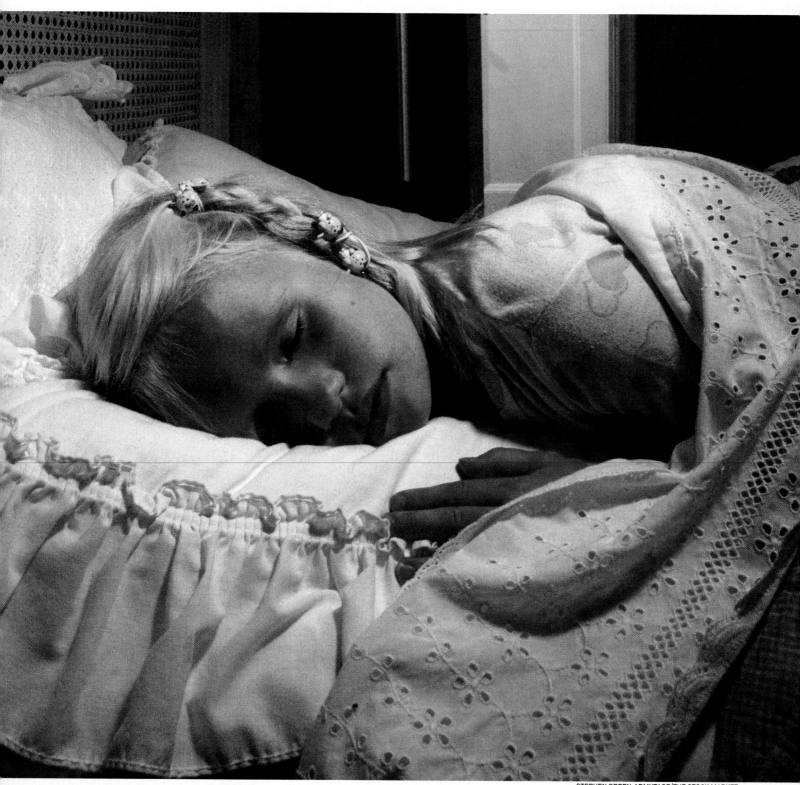

SHARON DAVIS

Why do I need sleep?

Bedtime. You may try to put it off, but it comes anyway. Every night you need sleep.

Scientists don't know exactly why sleep is necessary. One theory, which your own experience probably demonstrates, is that sleep refreshes tired people by restoring energy used up during the daytime.

Over the course of an ordinary night, you drift from drowsiness to light sleep to deep sleep. Your muscles relax. You breathe—and your heart beats—at a regular rate. After about 90 minutes, you enter Rapid Eye Movement, or REM, sleep. In this more active stage of sleep, your eyes dart back and forth under closed lids, and your breathing and heartbeat become irregular. You do most of your dreaming during this stage. You may switch from REM to non-REM sleep and back again several times a night.■

Why do I dream?

Chances are you entered a dream world last night. Today you probably can't remember what you dreamed about. Almost everyone dreams almost every night. According to some researchers, people often dream about ordinary events of daily life. Sometimes people have strange dreams and frightening nightmares. They may remember best the dreams they have right before waking.

Dreams occur mainly during REM sleep. The purpose of dreaming is uncertain. Some scientists think that dreams reveal a person's inner feelings. About 90 years ago, a doctor named Sigmund Freud suggested this theory in a famous book. Other scientists think dreams are a response to memories and to the sights and sounds a person experiences. According to a different theory, dreaming during REM sleep simply refreshes the brain by clearing overloaded circuits.■

How do bats rest

When people rest, they sit or lie down to "take a load off their feet." When bats rest, they put a load *on* their feet. Most bats coming home to roost hang upside down from their feet. They grasp a twig or find a toehold on the ceiling of a cave.

This might seem to be a batty way of resting, but it suits most kinds of bats. Like you, a bat has five toes on each foot. The bat can use its toes and pointed claws like hooks to suspend itself from a perch.

As soon as the bat is ready for action again, it releases its grip, spreads its wings, and swoops into flight. Among mammals, only the bat can truly fly.■

MERLIN D. TUTTLE, BAT CONSERVATION INTERNATIONAL

SHARON DAVIS

Why do some people sleepwalk

You're sleeping over at a friend's house. Late at night, the friend rises from bed, opens a door, and walks into another room. You wake up and call out, but get no reply. Your friend is sleepwalking, and may not see you or hear your voice. At first, you may not even realize your friend is asleep. Unlike the wandering girl above, most sleepwalkers move with their eyes open and their arms down.

A person who awakens for a short time while sleepwalking probably will have no memory of either the sleepwalking or the brief awakening.

Doctors believe they know *when* most people sleepwalk, but not *why*. Sleepwalking commonly occurs during passage from a deep to a lighter stage of non-REM sleep. Sleepwalking episodes usually don't last very long, and the sleepwalker usually doesn't go very far. Many children sometimes walk in their sleep. Most stop doing so as they grow older. Since sleepwalking seems to run in families, a child who sleepwalks may discover that some relatives once sleepwalked, too.■

10

Why do some animals hibernate?

Sleepy: It made a good name for one of the seven dwarfs who befriended Snow White, and it would make a good name for a common dormouse like the one above. A dormouse sleeps at least half the year. Scientists refer to its deep sleep as hibernation.

Some animals like the arctic fox remain active in cold weather. Others hibernate to survive harsh winters when food is scarce. During summer, they eat plenty of food, enough to build up large supplies of body fat. When they hibernate, they use little energy and require little food. They live mostly off their stored fat. Like the dormouse, many animals prepare for hibernation by lining their nest or burrow with twigs, grass, and leaves. The lining makes a warm, dry bed.■

Do fish sleep?

A diver swims among fish in their undersea world. With eyes wide open, the fish are resting. Their eyes always remain open because they have no eyelids to close.

Some fish seem to go into a state similar to sleep when they rest. Scientists don't know how closely the "sleep" of these fish resembles the sleep of other animals.

Certain kinds of fish rest while floating in water. Others snooze on the bottom, where they may hide among rocks, stand upright, or bury themselves in sand.■

11

Why does the wind blow?

A gentle breeze cools you. A steady wind lifts your kite. A sudden gust might blow your hat off. You have felt many kinds of winds, and you know that wind is moving air. The uneven heating of the earth's atmosphere sets the air in motion.

The sun warms the earth's surface, which in turn warms the atmosphere. Some parts of the earth receive direct rays from the sun all year and are always warm. Other places, like the North and South Poles, receive indirect rays, and the climate is colder.

Warm air weighs less than cool air. Because warm air is lighter, it rises. Cool air moves in to take the place of rising warm air. This movement of air is what makes the wind blow.

A windy day can mean fun. This man uses wind power for a fast ride on a sailboard. A windy day can also mean that wind turbines like those behind the sailboarder work well. Like giant pinwheels, wind turbines catch the breeze. Propellers turn, moving machinery that generates electricity. ■

JEFF DIVINE

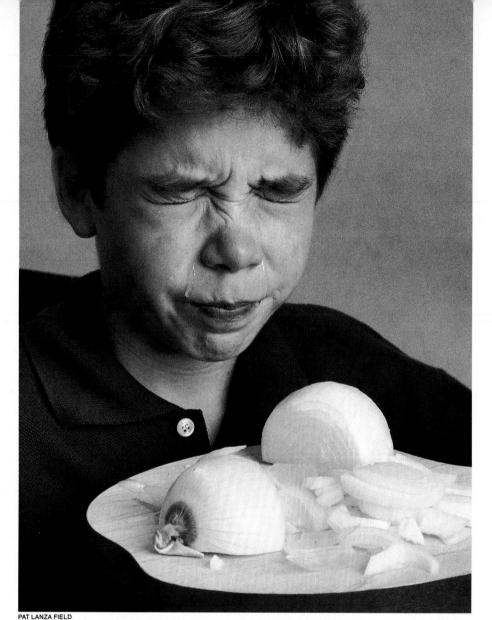

PAT LANZA FIELD

Why does slicing an onion make me cry?

It's invisible, but it brings tears to your eyes. What is it? A chemical released into the air from an onion when it is sliced. The chemical reacts with moisture on the surface of your eyes to form a weak solution of sulfuric acid. The acid stings. In self-defense, your eyes produce tears that dilute the acid and wash it away.

When it comes to onions, if you want to cut out the crying, try this: wear a pair of swim goggles. You may look a little silly in the kitchen, but your eyes won't feel the sting. Another tip for chopping without weeping: chill an onion before slicing it. Cold onions release smaller amounts of the chemical into the air. Some people say they cry less when they cut an onion under cold running water.■

Why do I sneeze?

Ah-ah-ah-CHOO! Something has irritated your nose, and you've just sneezed a powerful sneeze. Sometimes a sneeze can be so strong that the air rushes out at 100 miles an hour. That's faster than some hurricane winds!

Sneezing is your body's way of keeping your nose clear so you can breathe easily. Suppose dust enters and irritates your nose. Nerve endings inside your nose send a message to your brain to start the sneeze reflex. Like the woman in this picture, you take a deep breath automatically. Your lungs push air up and out. Finally, *achoo!* You sneeze out a cloud of air, water droplets, and, usually, the dust that started the trouble.

Can a breeze make you sneeze? Yes, especially if the breeze carries plant pollen that irritates your nose. The common cold may also cause you to do a lot of sneezing. Some people even sneeze when they look at the sun or at a bright light. Scientists think the light may irritate the optic nerve and trigger the sneeze.■

AL FRANCEKEVICH

STEVE NIEDORF/IMAGE BANK

Why does my mouth water?

Yum. Fresh cookies straight from the oven. It's enough to make your mouth water! Actually, the liquid that forms when your mouth waters contains more than water. The liquid is saliva and it contains chemicals called enzymes (EN-zimes).

Your brain activates the glands that produce saliva in your mouth. This can happen when you see, smell, eat, or even think about food. Saliva softens food in your mouth, making it easier for you to chew and swallow. Enzymes in saliva break large food molecules into smaller ones, helping you digest your food.

You may not find saliva an appetizing subject, but saliva improves your appetite. Food tastes best when it is mixed with saliva. Your taste buds wouldn't work as well in a dry mouth.

How much saliva does your body make? Every day, your glands produce more than a quart of it.■

DOWN TO EARTH

Why does the earth move around the sun?

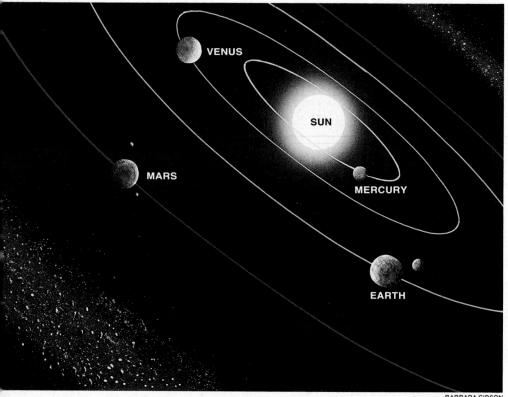

VENUS

SUN

MERCURY

MARS

EARTH

BARBARA GIBSON

For billions of years, the earth has been racing around the sun on a non-stop journey. An invisible force — gravity — holds our planet in its orbit.

Gravity pulls objects in the universe toward each other. An object that has a lot of mass has a stronger pull than an object that has less mass. The mass of an object is a measure of the total amount of material that it contains. Because the mass of the sun is enormous, its gravity is extremely powerful — powerful enough to keep the earth and all the other planets in our solar system in orbit around it.

The speeds at which the earth and the other planets move also play a part in holding them in orbit around the sun. If the earth were to travel any faster than it does, its speed would carry it into an orbit farther from the sun. If the earth were to travel more slowly than it does, gravity would pull it into an orbit closer to the sun.

In the diagram above, you can see that Mercury travels in an orbit closest to the sun. Mercury feels a stronger pull of gravity, and moves at a faster speed, than do planets that are farther from the sun. ■

Why does the earth seem still if it's really spinning?

High above an observatory in Hawaii, stars appear to streak around the sky in ever-widening circles. The trails they make look like a spectacular and mysterious kind of sky writing.

No matter how many times you gaze at the stars, you will never see a scene like this. A photographer used a special technique to capture in a single picture about two hours' worth of apparent changes in the positions of the stars. In the picture, the stars seem to move. In reality, the earth and the camera on it are moving, not the stars.

The earth rotates around an imaginary line, or axis, that extends from the North Pole to the South Pole. It takes nearly 24 hours for the earth to complete a full rotation, or spin, on the axis. The earth spins fast — about 1,000 miles an hour at the Equator. That's much faster than most jet airplanes travel. Yet you stand on this whirling planet and feel no motion.

Everything seems still because you are rotating at the same speed as all of the objects around you — buildings, land, rivers, oceans, even the air. Also, since a full rotation takes nearly 24 hours, your view of the positions of the sun and stars changes very slowly. As a result, you don't get a sense of any spinning motion.

Compare this to the feeling you have when you ride a merry-go-round. As your carousel horse goes round and round, you can tell you are spinning, in part because you move past the same spots over and over again during the few minutes of your ride.

Why don't you fly off the spinning earth? Gravity keeps you and everything around you, including the atmosphere, down to earth. ■

Why do stars shine and appear to twinkle?

When you wish upon a star, you're making your wish on a huge, glowing ball of gas. That's what a star is.

In this picture made in Arizona, stars shine as a comet streaks across the sky. Scientists believe that each star forms from a vast cloud of gas and dust floating in space. Over millions of years, the gas and dust contract. Gravity pulls surrounding matter to the developing star. The weight of a star's outer layers creates enormous pressure and extremely high temperatures near the center of the star.

In most stars, the pressure and high temperatures cause hydrogen to change into helium in a process known as nuclear fusion. Heat and light are released during this process. Result: a shining star.

Most stars shine for billions of years. Our sun—the star that is closest to the earth—has been shining for about five billion years. Scientists predict it will shine for five billion more.

On a clear night you might be able to see thousands of stars in the sky. Their light seems to flicker and blink. The stars do not really twinkle; they only appear to do so. Starlight passes through moving layers of air as it travels to earth. Some of these layers are more dense than others. Light rays bend when they travel from the stars through layers or currents of air of different densities. This gives the stars their twinkling appearance.■

FRANK ZULLO

Why is space dark if stars are shining?

When you stand on the earth and gaze into space, you may see a scene similar to the one on the left: a sky filled with shining stars. If you ever travel in space, like the astronaut above, you'll see a different, much darker scene. Who turned out the lights?

Light waves from the sun and other stars pass through the earth's atmosphere. Along the way, they strike air molecules and dust particles, and scatter. Some scattered light waves come toward the ground, making the sky look brighter.

In many regions of space, gas and dust barely exist. Light waves from stars find few gas molecules or dust particles to strike. As a result, most light waves do not scatter and do not shed light. In such regions of space, you may see nothing but darkness.■

Why is gravity much stronger on earth than in space?

"Chin up!" Schoolchildren in Iowa huff and puff and stretch and strain at a chinning bar (left). Gravity holds them down. To pull themselves up to the chinning bar, they struggle against the force of gravity with muscle power.

Gravity is not a problem for astronauts on a mission in space (above). One floats weightlessly inside a space shuttle cabin. Restraints hold her fellow astronauts in place.

Gravity acts like a strong, invisible glue. In space, gravity binds objects together. It holds the moon, the earth, and other planets in our solar system in their orbits.

Despite its awesome strength, gravity can be weak. Gravity between objects decreases as the distance between them increases. Also, objects with less material, or mass, exert a weaker force of gravity than objects with more mass. The mass of the earth is less than that of Jupiter, but more than that of the moon. So gravity on earth is weaker than it is on Jupiter, but stronger than it is on the moon.

In space, enormous distances separate the moon, the earth, Jupiter, and other objects of great mass. Across such distances, the force of gravity is extremely weak.

Gravity affects the weight of an object, not its mass. A man who weighs 200 pounds on earth would weigh approximately 500 pounds on Jupiter but only about 32 pounds on the moon. His mass wouldn't change. In the even weaker gravity of space, the same man would weigh so little he would float.■

No Body Like Your Body

Why do I blink?

Think about blinking. Every day, you blink thousands of times. Each blink lasts a fraction of a second. But those blinks add up. On an average day, your eyes might be closed in blinking for more than 30 minutes!

You blink automatically. When you do, your lashes and lids drop quickly over your eyes. The lashes help keep out dirt and dust. The lids bathe your eyes with a small amount of tears that prevent your eyes from drying out.

When something moves quickly near your face, your eyes blink in self-defense. If a small object, like a gnat, flies into an eye, you'll blink and produce tears to flush the object out. If a larger object, like a baseball, speeds toward your face, you'll blink. But when that happens, blinking won't help. Fortunately, you'll probably duck too!■

Why do I see spots after looking at a bright light?

You don't have to be a magician to make spots appear before your eyes. If you look at a bright light and then look away, you may see a spot in space. If you shift your eyes in another direction, the spot shifts, too. In fact, you may continue to see the spot even with your eyes closed!

The spot is an optical illusion known as an afterimage. It occurs, some scientists believe, when bright light triggers a chemical change in your eyes. As a result of the change, a substance called visual purple, or rhodopsin (row-DAHP-sun), is temporarily bleached from your eyes. In a process scientists don't yet fully understand, the bleaching causes the afterimage to appear in your sight as a spot. After a few seconds, rhodopsin levels usually return to normal, and your vision once again is spotless.■

Why do my veins look purple if my blood is red?

Your blood is bright red—but not all the time. It varies in color from scarlet to deep purplish red, depending on where it is as it flows continuously through your body.

A substance in red blood cells called hemoglobin (HEE-muh-glow-bin) gives blood its color. When hemoglobin combines with oxygen, a chemical change takes place, and blood turns bright red. This happens when blood passes through your lungs, where it picks up oxygen from inhaled air.

From the lungs, blood travels to your heart, and then through large blood vessels called arteries to the rest of your body. On the way, it exchanges oxygen for carbon dioxide, a waste product made in cells. When hemoglobin gives up oxygen, blood turns from red to purplish red.

It is blood of this color that travels back to your heart through blood vessels called veins. In some places on your body, veins lie near the surface of the skin. Look at the back of your hand or the inside of your wrist or elbow. Depending on the color of your skin, the blood in the veins underneath may appear purplish or bluish.■

Why do parts of my body "fall asleep"?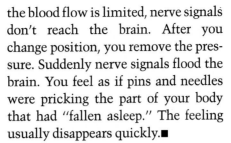

You've been sitting with your legs crossed. You get up and take a step, but stumble. You haven't tripped over anything. One of your legs has "fallen asleep." Until it "wakes up," you may limp or hop around on the other leg.

The tingling sensation you may feel when a part of your body "falls asleep" usually occurs because you have been pressing a nerve between a bone and another hard object. This might happen when you sit on a hard chair and cross your legs, or dangle your arm from the back of the chair, and remain in the same position for a while. In such a position, you may also compress a blood vessel.

When a nerve is under pressure or the blood flow is limited, nerve signals don't reach the brain. After you change position, you remove the pressure. Suddenly nerve signals flood the brain. You feel as if pins and needles were pricking the part of your body that had "fallen asleep." The feeling usually disappears quickly.■

SHARON DAVIS (ALL)

Why do my ears sometimes ring?

Rrrring. This boy hears bells. But no bells are jingling within earshot. The noise comes from inside his head.

At some time or other, nearly everyone experiences a mild form of tinnitus (TIN-uh-tuhs), or noise in the ear. Often the noise sounds like ringing. Sometimes it sounds like buzzing, popping, clicking, or roaring. A person with an ear infection—or a person who is worried and nervous—might experience tinnitus. Scientists don't know what causes all the sounds, but they do know some of them come from a bad connection between the inner ear and the brain.

In your inner ear are hair cells that help you hear sounds around you. If the hair cells are damaged, they can trigger nerves to send false messages to the brain. Result: You hear sounds that come from inside your head.

Do you like to play loud music frequently? Turning down the volume will eliminate one cause of damage to the hair cells: loud noises.■

What causes high waves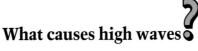

A high ocean wave provides a fast ride for a surfer in Hawaii. The wave motion carrying him along started thousands of miles away near Alaska and the Soviet Union.

Most waves form when wind blows across the ocean. The particles of water in a wave travel in circles, without much forward movement. But the shape of a wave can travel great distances across the ocean. The size of a wave depends on how hard, how long, and how far across the ocean the wind blows. Prevailing winds blow steadily from the same direction. When strong prevailing winds blow for a long time over great stretches of ocean, the result is high waves at sea.

As waves approach shore, the ocean floor slows them down. Where the ocean floor rises and the water becomes more shallow, waves usually get higher and eventually break. The highest waves form where the bottom of the ocean rises in a steep slope. You can find high waves at some beaches in Hawaii, in California, and in many other parts of the world.■

JEFF DIVINE

Why do "tidal waves" occur

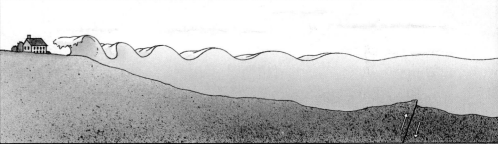

BARBARA GIBSON

"Tidal waves" have nothing to do with tides. The huge waves—more properly called tsunamis (tsoo-NAH-meez)—are caused by sudden movements of the earth under the ocean.

In the diagram, an earthquake moves the ocean floor. The movement generates a tsunami. From land, one of the first signs of a tsunami can be a giant vacuum effect that sucks water away from the coast. Then the tsunami slams into the shore. The destructive wave forms a rushing wall of water that can sweep away cars, trains, houses, or even an entire town.

Normally, after a pause of about 15 minutes or more, another wave roars ashore. The waves may continue coming for three to four hours or more. In 1755, according to historical records, a series of such waves battered the European coast for a week!

"Tsunami," from the Japanese, can be translated as "harbor wave." From the Pacific Tsunami Warning Center in Hawaii, scientists issue advance notice of approaching tsunamis. People in danger can go to safe ground.■

Why do I get thirsty?

Warm day, cool water: This boy is enjoying both. He drinks to quench his thirst, probably unaware that he is replacing water lost through sweating.

Your body needs a certain amount of water. One job of the hypothalamus (hy-poh-THALL-uh-muss), a portion of your brain, is to regulate that amount. On a normal day you lose about two quarts of water in the form of sweat, wastes, and the moisture in breath. Your hypothalamus triggers a feeling of thirst, and you drink to replace the lost water. If you eat a lot of salty foods, you may take in more salt than your body requires. Once again, you feel thirsty. Your body is signaling you to wash out the extra salt with water.■

© 1988 LAWRENCE MIGDALE

How can camels go without drinking for long periods of time?

The sun beats down on people and camels trekking across the desert (top). In the extreme heat of a desert summer, camels can go without drinking water for several days. In winter, when the desert climate becomes cooler, camels can survive without drinking for weeks.

The camel accomplishes this remarkable feat by doing three things: getting moisture from desert plants that it eats, drinking a lot of water

26

whenever it has the chance, and keeping the water that it drinks inside its body for a long time.

At a watering hole like the one above, a camel can, in 10 minutes, slurp up 30 gallons of water. That's almost 500 cupfuls!

A camel loses the water in its body slowly. It doesn't sweat easily, and its droppings are dry.

The camel stores water in its body tissues. The animal's hump provides storage space for fat, not water. When the camel finds enough to eat, the amount of fat increases and the hump swells. When the camel finds less food, it lives off the fat. Then the hump shrinks.■

Where does dew come from?

Even on a dry day, the air around you has some moisture in it. The moisture is in the form of a gas called water vapor. Warm air holds more water vapor than cold air does. On a warm day, water evaporates into the air. Then night falls. The air cools, and it can't hold as much water as it could during the day. On flower petals, on blades of grass, and on other cool surfaces, the water vapor condenses, or turns back into a liquid. Early the next morning, you can see the liquid glistening as dewdrops. After the sun rises, the air warms up and the dew evaporates.

You'll see dew most often in spring and fall. In summer, night air may remain warm and continue holding water vapor. At other times of the year, if the ground is cold enough, water vapor in the air can change directly into ice. Then you may see frost on the ground, not dew.■

27

Why does a dry road sometimes look wet on a hot day ?

A cool pool of water appears to be covering this road in Death Valley, California. No matter how fast or far the jogger runs, he won't get wet in the "pool" because it isn't there. The "rippling water" is a displaced image known as a mirage (muh-RAHJ).

Rays of light usually travel in a straight line, but under certain circumstances they bend. Here the light rays bend as they pass through layers of air that have different temperatures. They curve toward the cooler air.

In Death Valley, the sun bakes the desert floor. Heat rises from the ground, and warms the air directly above it. Light rays from the sky approach the warm air near the ground. Then the light rays bend upward toward cooler air. As a result, an image of the sky appears on the road. The image shimmers because the light rays pass through layers of air that are heated unevenly.

If the shimmering water doesn't exist, why does it appear in the picture? Mirages are actual light rays, and so they show up in photographs.■

Why can't people survive on salt water?

A thirsty sailor dreams of a frosty pitcher of fresh water. Unfortunately, drinking the ocean water that surrounds him would only increase his thirst, not quench it.

Your kidneys flush extra salt out of your body as waste. Ocean water is very salty. If you drank it, you'd take in much more salt than your body could get rid of. You would need more water to flush out the salt, but you couldn't get enough water from salt water to do the job. Soon the extra salt would make you sick.

A sea gull like the one perched on the raft has a salt disposal system that is different from yours. This bird expels extra salt through tubes along its beak. The tubes come in handy, since the sea gull often drinks salt water. ■

Hot Times

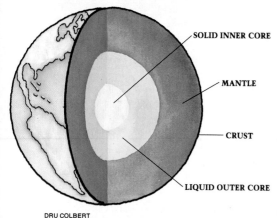

SOLID INNER CORE

MANTLE

CRUST

LIQUID OUTER CORE

DRU COLBERT

Why is it hot at the core of the earth?

Inner earth is full of mysteries. Scientists who study the earth tell us what may lie beneath the surface. They say that the earth consists of layers shown in the simplified diagram above: a thin crust, a thick mantle, and a core with a liquid outer section and a solid inner section. Scientists also think that the mantle and crust consist of rocky materials and the core consists mostly of iron. If you could measure the temperature of the core, you would find it hot—perhaps 9,000°F or higher!

Some scientists believe that, billions of years ago, materials that form the earth were distributed throughout the planet. As the earth grew larger, new materials piling up on the outside squeezed materials inside. Energy used in this process generated heat. Radioactive materials decayed, creating more heat. Eventually most of the iron in the earth melted and sank to the core. This created still more heat. Lighter, rocky materials rose to form the crust and mantle.

Temperatures in the mantle and in parts of the crust, though cooler than at the core, are still hot enough to melt rock under certain conditions. Volcanoes such as this one in Hawaii erupt where melted rock called magma rises to earth's surface. The magma originates in the upper mantle and crust.■

ROBERT W. MADDEN, N.G.S. STAFF

Why does blowing make some flames grow bigger and others go out?

Children blow on candle flames (right), and the flames go out. Another child blows on a campfire (above), and it grows bigger. How can the same action produce different results?

You need three ingredients to start a fire: heat, oxygen, and fuel. The fuel can be wood, a candle, certain chemicals, paper, or anything that burns. When you strike a match, friction produces heat. The heat causes oxygen in the air to combine with chemicals on the tip of the match to produce fire. You can put the fire out if you take away one of the three ingredients. You can make the fire bigger if you add more fuel or oxygen.

By blowing hard on the candle flames, the children remove the heat. That extinguishes the flames. By blowing gently on a larger flame, the camper adds oxygen to the fire. That makes the flames grow. If a strong wind blows, it might remove heat from the campfire. Then the campfire would go out just as the candles did. ■

SUSAN M. KLEMENS

JIM WRIGHT/NAWROCKI STOCK PHOTO

Why do things turn black when they burn?

As the logs in this campfire crackle and burn, they will turn black. A marshmallow, roasted too long on one side, will also blacken. Objects generally turn black when they burn, no matter what color they were before.

Most familiar substances that burn contain the element carbon. Many of these substances—such as logs, paper, certain kinds of cloth, and food—come from plants or animals. Atoms of carbon exist in all living things, usually in combination with other elements such as oxygen and hydrogen.

When a substance burns, molecules made up of carbon and other elements break apart. The elements then combine to produce new substances such as carbon dioxide and water vapor. Most of the new substances are gases that drift away into the atmosphere. Some of what remains is carbon that did not combine with other elements. The black color of the carbon makes the burned object look black.■

WARMING UP, COOLING DOWN

What is the difference between warm-blooded and cold-blooded animals?

TONY STONE WORLDWIDE/MASTERFILE

© GÜNTER ZIESLER—PETER ARNOLD, INC.

In the animal world, what is "cold" can also be warm. A cold-blooded animal like a tree frog (above) may become warm by sitting in the sun.

Many people would use the phrase "warm-blooded" to describe this lion and cub (right) and other mammals such as dogs, whales, and human beings. Warm-blooded creatures, which also include birds, generate their own body heat. They keep the heat inside with layers of feathers, fur, or fat. Some of them cool off by panting or sweating. In a warm-blooded animal, the brain regulates the body temperature, which does not change much in reaction to the outside temperature.

Cold-blooded animals include amphibians such as frogs, and reptiles such as snakes and lizards. A cold-blooded creature cannot regulate its body temperature from within. It soaks up warmth or coolness from its surroundings. If you pick up a cold-blooded animal that has been basking in the sun, you'll probably find that it feels anything but cold to the touch. It may feel even warmer than a warm-blooded creature!■

Hit the Heights

CHRIS NOBLE/ADVENTURE PHOTO

How do scientists measure the height of mountains?

Even if this climber were carrying a yardstick, he wouldn't be able to measure the mountain behind him (right). It's too tall. Researchers rely on a combination of mathematics and scientific instruments to calculate the height of mountains.

An old method of measuring a mountain's height involves barometers. Air pressure gradually decreases above sea level. With barometers, scientists can determine air pressure on a mountaintop. Since they already know what the air pressure is at different elevations, they can use the barometric readings to estimate the height of a mountain. Scientists must climb the mountain to use this method, and they must take many readings and average them to get good results.

A better method is triangulation. Scientists measure an imaginary straight line, called a baseline, across part of the mountain. Then, from a point at each end of the baseline, they view the summit against the horizon through a telescope-like instrument. They draw two imaginary lines from each end of the baseline to the mountaintop to create an imaginary triangle. They measure the angles of the triangle and use a type of mathematics called geometry to calculate the peak's height. This method works only when scientists can see the top of the mountain.

The newest, most accurate method relies on satellites to beam radio signals to stations on or near a mountain. Scientists base calculations of the mountain's height on the time it takes the signals to reach the stations. A satellite's orbit serves as a baseline in the sky for the measurements. ■

Why is it hard to breathe on high mountains?

Secured by strong ropes, a mountain climber scales a tall peak in Alaska. The air he is breathing contains less oxygen than does the air down below.

Everybody needs oxygen. When you breathe, you take this gas into your body from surrounding air. At high elevations, air is thin because there are fewer molecules of all the gases in the atmosphere, including oxygen.

The shortage of oxygen makes it hard to breathe at great heights.

People unaccustomed to the high life may get headaches and feel dizzy and nauseated. Some carry an oxygen supply along on climbing trips. But people who spend a lifetime on high mountains breathe more easily. They develop extra red blood cells that help supply the oxygen they need. ■

PAT MORROW/FIRST LIGHT, TORONTO

36

BEASTLY BEHAVIOR

© JOHN W. WARDEN

How can some animals climb steep mountains?

"Have grip, will travel," might be a good motto for the young mountain sheep (left) and for the mountain goats (below). Whether they are going up a steep slope or over a slick surface, these surefooted creatures can leap and climb easily.

Tough, rubbery hooves split into two toes give the animals their firm grip. The toes can spread wide or draw together to grasp slick rock. Under each toe is a rough pad. The pads hold on to rock—even rock that is wet or icy. Hard edges and pointed tips on the hooves help the animals cling to tiny cracks in rock.

Mountain goats can perform tricky balancing acts. Muscular bodies and short legs help them stay steady on steep cliffs. Occasionally one of them may stumble and fall. Unless it is injured, it will get right back up and start climbing again. ■

DES & JEN BARTLETT

Why do kangaroos "box"?

In a quiet field in Australia, two rivals circle each other cautiously. One punches and kicks. The other lashes out to push the kicker away. They bite and claw, or try to wrestle each other to the ground. The "boxing match" rages . . . but not for long. Within minutes, the struggle ends. Neither fighter is seriously hurt.

What's all the fuss about? These boxers are male kangaroos, and they are probably fighting over a female. Male kangaroos also fight to protect their territory against intruders. Some observers think they sometimes fight just for fun! When kangaroos finish a fierce fight, they become gentle animals once again. They often go back to grazing peacefully, side by side. ■

DAVID C. FRITTS

Why do raccoons "wash" their food?

Some people dunk bread into gravy, or doughnuts into coffee. Like them, a raccoon sometimes moistens its food.

A raccoon will eat almost anything: frogs, berries, insects, mice—or the contents of your garbage can! If a raccoon finds something tasty to eat, it may dip the food into a nearby stream or even into a water dish. Then it may rub the food underwater with its paws. The raccoon appears to be washing the food, and people once believed it was doing exactly that. Today, however, scientists have other explanations for the animal's behavior.

According to one theory, the raccoon dips its food out of habit, because it finds so much of its food in the water. Another theory is that the animal is examining its food. The raccoon has a keen sense of touch, which may improve when its paws are wet.

Raccoons don't always dunk before dining. If no water is nearby when they find something to eat, they may gobble up the meal on the spot.■

How do webbed feet help ducks swim?

If you wanted to move through the water in a canoe, you would use a paddle. When a duck swims, it uses objects that work like paddles—its own two feet.

Take a look at these ducklings. Through the clear water, you can see the flaps of skin that form the ducklings' webbed feet. As a duck swims, it moves its legs backward and forward. On the forward stroke, the duck closes the toes, folding the web. In this position, the foot moves through water easily. On the backward stroke, the duck spreads the toes, stretching the web open. The foot then works like a wide paddle.

The duck moves by pushing its "paddles" against the water. To change direction, the duck paddles harder with one of its feet than it does with the other.

In water, a duck glides smoothly and gracefully. On land, a duck looks clumsy. Webbed feet combined with legs set wide apart at the back of its body make it "waddle like a duck." ■

J. SHERWOOD CHALMERS

Do flying squirrels and flying fish really fly?

Easy gliders, a flying squirrel (above) and a flying fish (right) sail through the air. Since neither animal has wings, neither can really fly. They travel through the air by gliding.

A flying squirrel has loose folds of skin on its sides, between its front and back legs. To glide, the animal leaps off a tall branch and extends its feet. The loose folds of skin stretch to form a kind of parachute. Then the animal can glide smoothly from tree to tree.

A flying fish doesn't have such a high launching pad. It takes off from water. To become airborne, the fish pushes hard with its powerful tail. Large fins spread wide and allow the fish to glide half the length of a football field or more before dropping back into the water. Some flying fish have leapt high enough to land on the decks of ships! ■

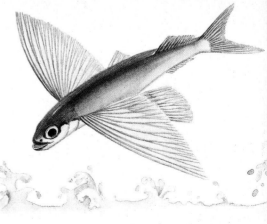

BIRUTA AKERBERGS-HANSEN

41

Why do I see myself in a mirror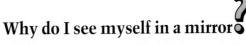

STEVE LEONARD/BLACK STAR

The Ugly Duckling, a fairy tale character, stared in amazement at the smooth surface of a stream. He saw the reflection of a beautiful swan. If something had rippled the calm waters, the "duckling" might not have discovered his true identity.

Some things, such as the sun and electricity, produce light. Nearly everything else reflects light. A rough surface—such as that of rippling water—scatters the light that it reflects in many directions. A smooth, shiny surface—such as that of calm and clear water—does not scatter light, but reflects it in one direction.

Most mirrors are made of glass with a thin backing of silver or aluminum. They have a smooth and shiny surface. So do other objects—a polished tabletop, for example, or a silvery tray.

RICHARD STEEDMAN/THE STOCK MARKET

Because such objects do not scatter light but reflect each ray of light in one direction only, they reflect images.

When you face a flat mirror like the one above, light rays do not scatter. They bounce from you to the mirror and back to you. You see an image of yourself that is the same size as you are and that seems to be as far behind the mirror as you are in front of it. Since this girl (above) sits between facing mirrors, she can also see reflections of her reflection.

A mirror sends light from one side of an object back to the same side. As a result, the image seems reversed. Take a look at the signs on this busy New York City street (left). They're not really reversed. They simply appear to be so because they are reflected in the windows of a building. ∎

Why can't people see ultraviolet light?

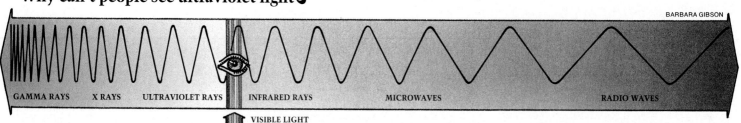

GAMMA RAYS   X RAYS   ULTRAVIOLET RAYS   INFRARED RAYS   MICROWAVES   RADIO WAVES

VISIBLE LIGHT

A child holds a handful of crayons (below left). The lighting changes (below right), and the crayons take on a mysterious glow. The glow you see is caused by rays of ultraviolet light—more properly called ultraviolet radiation—that you can't see.

The diagram (above) shows wavelengths in the electromagnetic spectrum. You can see only wavelengths of visible light. Their colors are the colors of the rainbow: red, orange, yellow, green, blue, and violet. Beyond red is infrared. Beyond violet is ultraviolet. The human eye cannot see infrared wavelengths because they are too long, nor ultraviolet wavelengths because they are too short. Although ultraviolet wavelengths are invisible to people, researchers think some insects can see them.

In the picture of glowing crayons, ultraviolet radiation comes from an electric lamp with a black-light bulb. Ultraviolet rays from the lamp fall on the crayons, which are fluorescent (floor-ESS-uhnt). Atoms in the crayons absorb energy from the rays and give it off at a longer, visible wavelength. You see the result as the glow of fluorescence. When the lamp is turned off, the glow disappears.

You might be surprised to learn that thousands of different objects fluoresce under light from an ultraviolet lamp. Besides fluorescent crayons, these objects include certain kinds of minerals and chemicals, some spoiled foods, even your teeth!■

Why does a hologram appear three-dimensional?

In a museum in Paris, France, a boy gazes at an image of a woman talking on the telephone. The image is a hologram. It revolves inside a plastic cylinder. As it does, the boy can see various sides of the image, just as he could see various sides of a real person.

Like photographs, holograms are pictures. Unlike photographs, holograms appear three-dimensional. Photographs are generally made with ordinary white light. Holograms are made with laser light.

All light travels in waves. Waves of ordinary white light, such as sunlight, do not have the same length and do not travel together. Waves of light from a laser *do* have the same length and *do* travel together.

In holography, a beam of laser light is split into two beams. They travel through an arrangement of lenses and mirrors. One beam heads toward the object to be holographed. The other beam travels toward holographic film. The beams meet to form a pattern of light that contains information about the size, shape, and depth of the object. This pattern of light is recorded on the film. Light shining on the developed film is bent and focused by this pattern into a three-dimensional image of the object.

The revolving cylinder hologram pictured here is one type of integral (IN-tuh-gruhl), or movie, hologram. Most holograms are not cylindrical. Some can be displayed like paintings on a wall or used in a book. You may have seen holograms on stickers, on credit cards, or even on the cover of NATIONAL GEOGRAPHIC magazine.■

Skin Deep

Why does skin wrinkle?

When this old woman was a young girl, her skin was soft and smooth. Now, like many aging people, she has skin that folds in deep creases. Her wrinkles were caused in part by the process of growing older.

As people age, their skin slowly changes. It loses its elasticity, or stretchability. The cushion of fat cells beneath the skin also thins out. Without this padding, skin droops.

As people continue growing older, their bones may change. If this happens, people who already have some wrinkles may see more develop.

Aging causes wrinkles, but exposure to sunlight and to harsh weather can make wrinkles worse and bring them at an earlier age. The deepest wrinkles may appear on the skin of a person's face, neck, hands, and arms. That's because those parts of the body are usually exposed to the sun and wind most often.■

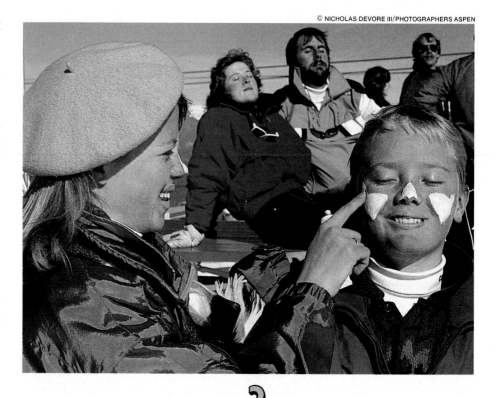

© NICHOLAS DEVORE III/PHOTOGRAPHERS ASPEN

© 1988 PAUL SLAUGHTER

Why does sunlight darken skin?

Winter's here! These children (above) take a break from skiing.

Summer's here! This boy (right) rides his skimboard in ocean surf.

Winter or summer, ultraviolet rays from the sun can cause your skin to darken. Certain skin cells produce a dark pigment, or coloring material, called melanin (MEL-uh-nun). This substance gives your skin its natural color. In sunlight, cells produce more melanin. The pigment absorbs ultraviolet rays, which can burn your skin. Melanin travels from cells where it is produced to nearby cells, and darkens them. In a few days, these cells reach the skin's surface. Result: you look darker than before.

Many people think exposure to sunshine is good for your health. Small amounts of ultraviolet rays do help your body produce vitamin D, which it needs for strong bones. But too much ultraviolet radiation at once can give you a sunburn. People who spend a lot of time in the sun over many years risk developing skin cancer.

Can you enjoy the sun and still protect your skin? Many experts believe you can, if you apply a sunscreen or a sunblock, like the zinc oxide preparation the children above are using.■

PETER READ MILLER/ SPORTS ILLUSTRATED

© NANCY BROWN

Why do I itch?

Can't stop scratching? Maybe an insect bit your arm or leg. Or perhaps you brushed against poison ivy. Now your skin is irritated. Your body reacts by releasing certain chemicals at the spot where you were bitten by a bug or brushed by a weed. The chemicals interact with nerves in your skin. The nerves send itch signals through your spinal cord to your brain, and you feel like scratching.

If you eat food you are allergic to, your body may respond in a similar way, releasing chemicals that interact with nerves in your skin. Once again, you've got that itchy feeling—and perhaps a rash as well.■

SHARON DAVIS

Why do some people have freckles?

Some people say freckles show you've been kissed by the sun. In fact, freckles occur because some cells in your skin produce more of a pigment called melanin than other cells do. Usually, melanin colors skin evenly. But when groups of cells that produce melanin at a normal rate surround groups of more active melanin-producing cells, the dark pigment appears in spots.

People often find themselves speckled with freckles after they spend time in the sun. The freckles emerge when the body produces melanin to absorb harmful sun rays, and the melanin develops in clusters.■

Why do I sweat?

It's an exciting game. This basketball player has been leaping high and running fast. Sweat covers his body. If he keeps exercising hard, he may soon look as if he had just stepped out of the shower. Sweating is a kind of shower, but the wetness comes from *inside*.

A part of your brain keeps track of the temperature inside your body. It also receives messages from nerves in your skin about changes in the outside temperature. When you use a lot of energy, when the weather is very hot, or even when you are frightened or under stress, your body may heat up. Then your brain will react to cool you down.

More than two million tiny coiled tubes called sweat glands lie in your skin. At a signal from the brain, these glands squirt watery fluid—sweat—onto the surface of your skin. From there, sweat evaporates, carrying away heat and cooling your body. So the next time you sweat, don't sweat it! Your body is keeping its cool.■

TOM BRAKEFIELD

How does an elephant use its trunk for drinking?

Part drinking straw, part water pistol: that's an elephant's trunk. Elephants slurp a lot of water. An adult elephant may drink more than enough each day to fill a bathtub! To take a drink, an elephant sucks water into its trunk. Then it curls the trunk into its mouth and squirts the water inside. Sometimes, after the elephant has had enough to drink, it fills the trunk again and squirts the water onto its back. The shower helps the animal stay cool, and keeps its sensitive skin from drying out.

An elephant also uses its trunk to gather food—anything from big branches to tiny nuts and berries. In dry weather, the animal digs for water with its trunk. It breathes through nostrils at the end of the trunk. And it touches other elephants with its trunk in a form of friendly greeting.■

Why do hornets sting?

© HANS PFLETSCHINGER—PETER ARNOLD, INC.

"Do not disturb." If a hornet could hang such a sign near its nest, it might never need to sting again.

Hornets sting mainly to defend their nests. They construct large nests of paper by gathering bits of wood, which they chew into a pulp. Unfortunately, the insects often build the nests near houses, where people find them a bother. If people touch a nest, by accident or on purpose, they might get stung. Hornets even take up guard duty at the entrance to their nests and sting other insects that try to come inside! Away from its nest, a hornet seldom stings unless it is disturbed.

Only female hornets can sting. They inject poison when they do. The sting can be painful, but it bothers some people less than it does others.■

Why does a snake stick out its tongue?

It isn't bad manners for a snake to stick out its tongue. It's a good move.

In the roof of a snake's mouth are two small holes lined with nerve endings. A snake sticks out its tongue and picks up microscopic particles from the air and ground. Then the snake places the two tips of its forked tongue into the holes. Scientists think the nerve endings in the holes carry messages about taste and smell from the particles to the snake's brain. The messages help the snake identify objects all around it, and to track prey.

If the sight of a snake sticking out its tongue startles you, remember this: a poisonous snake injects its poison through its bite, not its tongue. The tongue, darting in and out, may look dangerous, but it is harmless.■

Why are some animals born as runts?

Four Saint Bernard puppies snuggle next to each other for a snooze (below). One is smaller than the others. You might think it was much younger, but all four belong to a single litter born to the same mother at the same time. The littlest puppy is a runt.

Veterinarians tell us that animals develop as runts before birth. Some unborn animals have defects that prevent them from growing normally. Other unborn animals have no defects but, while still inside their mother, share nourishment with too many littermates. One of them may not receive enough nourishment, and be born as a runt. In rare cases, a mother may bear one litter that results from two matings. Newborns from the second mating could be runts because they had less time to grow than did their littermates from the first mating. Runts usually need special care, and even then they may not survive. ■

Why are albino animals white?

Like a white shadow, Goolara, an albino (al-BY-noh) koala, clings to its mother's back (right). The animals live in California's San Diego Zoo.

Like most albinos, Goolara inherited its white coloring from parents with normal coloring. Tiny particles called genes are found in all of an animal's cells. Genes form a code that makes an animal develop in a certain way. Some of the genes determine color. Animal parents with normal coloring may also carry a "no color" gene. If an animal inherits one "no color" gene from each parent, it develops as an albino. This happens only rarely.

In true albino animals like Goolara, the presence of the "no color" genes means that cells do not produce the dark skin pigment called melanin. Without melanin, albino animals are totally white. Mammals have white fur, birds have white feathers, and snakes have white scales.

Zookeepers take good care of Goolara. In the wild, albino animals have a harder time. Their coloring makes them easy targets for enemies. Most also have poor eyesight, which makes it hard for them to find food. ■

ON THE RISE

Why do helium and hot-air balloons rise?

GARY GREENE

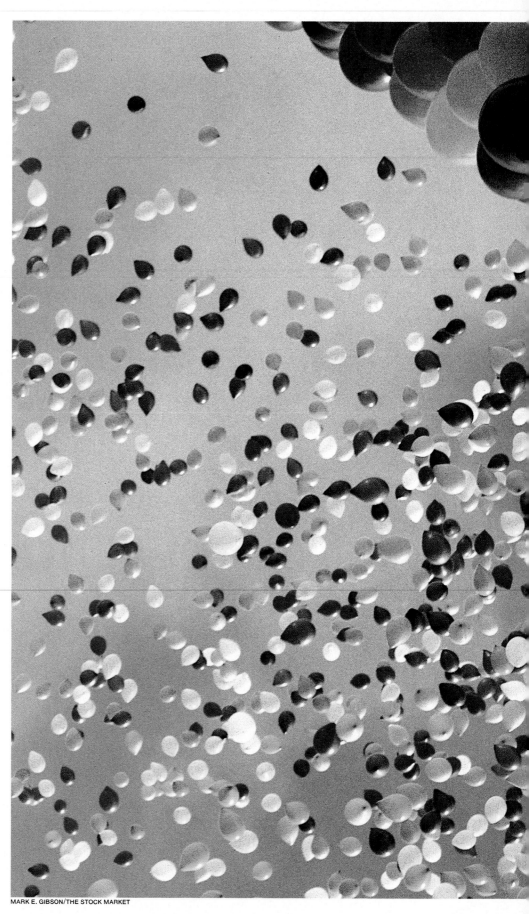

Hundreds of small helium-filled balloons dot the sky with color (right). A single large hot-air balloon reaches the end of its rope (above). It carries passengers aloft on a demonstration ride.

Warm air weighs less than cool air and tends to rise. Pilots heat air inside hot-air balloons with gas burners. Once heated, the air inside becomes lighter than the surrounding air. The balloons take off. Pilots land the balloons by letting heated air escape gradually or by controlling the heat that comes from the burners.

Helium balloons rise because the gas that fills them is lighter than air. They descend when the gas escapes.

If you release helium balloons outdoors, select your launch site carefully. Those that land in the ocean may endanger sea animals. The creatures may eat them or become tangled in their strings.■

MARK E. GIBSON/THE STOCK MARKET

What puts the "pop" in soda pop?

Every time you open a can of soda pop, you provide an escape route for carbon dioxide gas. *Whooosh!* The gas may burst forth in a spray of bubbles.

Carbon dioxide is a colorless gas that is present in the air around you. Manufacturers make soda pop by mixing flavored syrup, water, and other ingredients, and adding carbon dioxide under pressure. They seal the mixture, still under pressure, in containers. Some of the carbon dioxide is trapped in the empty space at the top of each container. Pressure forces the rest of the gas to dissolve in the liquid.

When you open the container of soda, you release the pressure. Gas escapes from the space at the top and from the liquid, too. If you set the open container aside, eventually your soda pop will lose all its "pop." ■

Why does bread dough rise?

A baker pulls loaves of warm, crusty bread from a brick oven (right). Even before the bread was baked, the dough that forms each loaf had puffed up.

Bakers make bread dough from ingredients that usually include water, flour, and sugar. They add another substance called yeast, which is actually a collection of thousands of single-celled organisms. The organisms consume the sugar as well as starch from the flour. They begin to multiply and produce carbon dioxide gas. Carbon dioxide—the same gas that is used in soda pop (left)— appears as bubbles in the bread dough. These bubbles make the dough expand and rise. ■

FIELD DAY

E.R. DEGGINGER

Why does a firefly light up?

The firefly is a shining example of insect communication. When the firefly shines its light, it tells other fireflies what species, or kind, of firefly it is and that it is ready to mate.

The firefly has a tiny light organ in the tip of its abdomen, or belly. It blinks the light to attract mates. Each species of firefly blinks in a different pattern. Patterns may vary in many ways — in the number of blinks, in the time between blinks, or in the color of the light. A firefly usually responds only to the pattern of its own species.

When a male firefly flutters over an area where it is likely to find females, it begins flashing its light pattern. A female of the same species blinks back in a pattern the male recognizes. The two continue exchanging flashes until the male lands near its new mate.

Other kinds of insects can glow, but only the firefly can flash a light on and off in code.■

Why do some flowers have fragrance?

"Stop and smell the flowers." That's good advice, as long as the flowers you sniff are bee-free. The same sweet smell that attracts you to certain flowers also attracts bees and other insects.

Certain cells in flowers produce chemicals that create fragrance. Petals, nectar, or other parts of the flower contain the cells. Different flower fragrances attract different insects. Moths flutter to honeysuckle and other flowers with a sweet scent. Beetles head for flowers such as the magnolia that have a fruity fragrance. Flies go for the foul odor of skunk cabbage.

Insects visit the flowering plants to find nectar to drink. At each flower, pollen clings to an insect's body. Insects carry the pollen to other plants. Plants need pollen to make seeds.

Sweet-smelling oils from the petals of certain flowers are used in some perfumes. Dab on a few drops of one of these perfumes, and you may come out "smelling like a rose."■

LON E. LAUBER/WEST STOCK, INC.

Why is grass green?

Roses are red, violets are blue . . . and grass is green. Chemicals called pigments give plants their colors. The pigment that tints grass green is called chlorophyll (KLOR-uh-fill).

Together with sunlight, chlorophyll enables plants to change carbon dioxide from the air and hydrogen from water in the soil into plant food. Many kinds of animals, including the fawn in this picture, eat plants. Others eat animals that eat plants. In this way, most animals benefit from the plant food that chlorophyll helps create.■

LEONARD LEE RUE III

WHAT'S IN A NAME?

KEN BALCOMB

FLIP NICKLIN: NICKLIN & ASSOCIATES

How did the killer whale get its name?

Sometimes killer whales seem playful. Like the one above, they breach—leap out of the water—and splash back down, possibly just for fun.

Sometimes killer whales seem gentle. A killer whale in a California sea park shows a friendly curiosity toward a young visitor (left).

Sometimes, however, killer whales are fierce. They have huge appetites. To satisfy their hunger, they zoom through the water after prey. They hunt fish, squid, octopuses, seabirds, seals, and dolphins. Banding together in packs, they even go after other kinds of whales that are bigger than they are! These aggressive hunting habits, many scientists now believe, earned the killer whale its name.■

Is the bald eagle really bald

"No!" The bald eagle below appears to be screeching the answer. As you can see, it's not bald at all; its head is covered with white feathers.

How did the name come about? Today's bald eagle, in earlier times, was sometimes called the "bald-headed eagle." In those days, the word "bald" meant "white" or "streaked with white"—a good description of the adult bird's white-topped head. Over the years, the term "bald-headed" was shortened to "bald." People today use the word "bald" to mean "hairless."

Many people think the name suits the bird. The contrast between the white feathers that cover its head and the dark feathers that cover most of the rest of its body can make the eagle appear bald from a distance. ■

Why do people call baby kangaroos "joeys"

If you find yourself at a loss for words, borrow some. Throughout history, people from different countries have borrowed words from other languages. When British settlers arrived in Australia, they heard Aborigines, or native Australians, calling a baby kangaroo a "joey." The British settlers borrowed the word, and spelled it the way it sounded in English. ■

SHINY TREASURES

Why are some gems more valuable than others?

All that glitters definitely is *not* gold. Emeralds, pearls, and diamonds glitter on a crown (right) formerly worn by a Middle Eastern monarch. Spanish noblewomen may have worn this necklace of diamonds and emeralds (above, far right), and a French queen may have owned these diamond earrings (below, far right). Whatever their history, the gems still dazzle.

Some gems develop from living things. Pearls, for example, can form inside shellfish such as oysters and mussels. Most gems, however, are minerals taken from earth's crust. Expert cutting and polishing make them catch the light and shine.

Sand and soil contain grains of minerals, but these minerals aren't valuable gems. The characteristics that set a valuable gem apart from an ordinary mineral include beauty, durability, rarity, and popularity.

People all over the world see beauty in diamonds, emeralds, sapphires, and rubies. People may admire the glittering surfaces of some of these valuable gems or the blazing color of others. Many prized gems are also durable. They can resist scratching or shattering, and may last for centuries.

The rarer a gem is, the greater its value. Even though red rubies and blue sapphires come from the same kind of mineral, corundum (kuh-RUN-dum), rubies generally are more highly prized than sapphires because there are fewer of them. Many valuable gems exist only in small quantities or only in certain parts of the world.

A gem that is beautiful, durable, and rare still may not be valuable unless it is also in fashion. Gems may gain or lose popularity and value depending on the fashion tastes of the times. ■

Why do some coins have ridges on the edges?

Take a close look at the contents of your piggy bank. You'll see vertical ridges, called reeding, on the edges of all your United States coins except pennies and nickels.

Many coins once contained gold and silver. Long ago, lawbreakers shaved bits of the precious metals off the edges of coins. They spent the coins, and sold the shavings. Some countries began marking or decorating the edges of coins to make it easy to spot those that had been shaved.

The United States Mint continued the practice of marking or decorating the edges of coins when it began manufacturing money about 200 years ago. Eventually the Mint put reeding on the edges of dimes, quarters, half dollars, and most gold and silver dollars. Today the Mint no longer makes gold or silver coins for general circulation, but it still puts reeded edges on coins. ■

NATIONAL GEOGRAPHIC PHOTOGRAPHER
JOSEPH H. BAILEY (ABOVE, BOTH)

THE COLD FACTS

Why do I shiver when I'm cold?

It's freezing! You've been playing outside for hours on a cold winter day. Suddenly you start to shiver. Don't be shook up. Your body is shaking to keep you warm.

Nerve endings in your skin sense any drop in the temperature. They send an alarm signal to a part of your brain called the hypothalamus (hy-poh-THALL-uh-muss). Your hypothalamus then transmits a message to the large muscles in your body: "Get moving!" In response, the muscles tighten and relax over and over again. These muscle movements—shivering—produce heat that helps warm your body. When you shiver, muscles may contract and relax as often as 20 times a second!■

Why do fingers and toes sometimes hurt in cold weather?

"Cold hands, warm heart." So goes an old saying about people, and scientists can tell you that it's true. Here's why.

Blood vessels carry blood throughout your body. When your body is exposed to the cold, it saves heat for your heart and other important organs by closing off small blood vessels in your fingers and toes.

Blood stops flowing freely through these blood vessels. As a result, the skin on your fingers and toes cools and you may feel pain. You may also find it hard to move your fingers and toes.

What can you do? Go inside. As you warm up, the small blood vessels reopen and blood returns to fingers and toes. You may feel a tingling sensation at first, but it will soon disappear.■

Why do I get goose bumps?

They cover the flesh of a plucked goose, and they can cover your skin, too. They are goose bumps, little bulges that can appear on your skin when you feel cold.

Hair covers most of your body. Next time you get goose bumps, look closely at your skin. You'll see the body hair standing up. Beneath your skin, the root of each hair is enclosed in a covering called a follicle (FAHL-ih-kuhl). When you are cold, a muscle connected to the follicle contracts. That makes the hair stand up and the skin bulge in a goose bump.

Fear may produce goose bumps, too. That's probably why some people say that their hair "stands on end" after a "hair-raising" experience.■

Why do snowballs sometimes fall apart❓

After a snowfall, you rush outside, scoop up some snow, and mold it with your hands. The usual result is a snowball. Sometimes, however, you end up with a handful of loose powder. The snow won't hold together because the weather is too cold.

As you pack snow together to make a snowball, the pressure of your hands causes some snow to melt. When you remove the pressure, the melted snow quickly refreezes and sticks together. Sometimes the weather is so cold that the pressure of your hands is not enough to melt the snow. That's when you're left with snow powder instead of snowball power.■

DRU COLBERT

65

What causes a snow avalanche?

Take a steep slope, add a lot of snow, and you have the makings of an avalanche. It doesn't take much to send the snow roaring downhill.

A change in temperature can trigger an avalanche. So can pressure from the weight of a single skier, or vibrations from the slamming of a car door. Many avalanches start under the weight of newly fallen snow. If the ground on a slope is rough, snow may cling to it. But if the ground is smooth, snow slides more easily. Then an avalanche is more likely to happen.

A big avalanche can thunder down the side of a mountain at speeds of up to 200 miles an hour. The rushing snow can flatten houses and toss cars into the air. For safety's sake, workers sometimes start an avalanche on purpose. They use explosives to blast snow loose before skiers arrive.

In the western United States, an estimated 100,000 avalanches occur every year. This avalanche in Utah (left) is one of them. In Alaska, the capital city of Juneau (JOO-noh) sits at the foot of a steep mountain (above). Some experts say an avalanche disaster is more likely to happen in Juneau than in any other city in the United States. ■

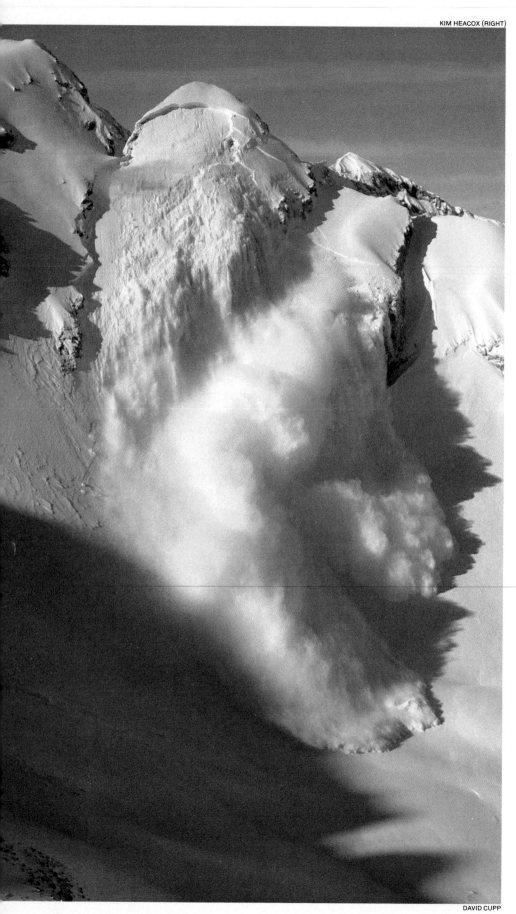

Why don't polar bears get cold?

Welcome to the polar bear's arctic world. In winter, temperatures may drop to -50°F. Blizzards sometimes fill the air with snow. Ice stays frozen all year long. Cold winds blow. Icebergs float in frigid seas.

Brrrrr. The Arctic doesn't sound like a cozy place to live. But polar bears live comfortably in their deep-freeze environment. They wear their cold weather protection all the time.

Almost hairless at birth, a polar bear cub soon grows thick hair that protects it from the cold. An adult polar bear stays snug in a heavy, two-layer coat. A dense inner layer of hair clings close to the body. An outer layer of long guard hair sheds water before it can turn into ice. Both layers help keep out arctic winds.

A polar bear's insulation includes, under the skin, a layer of fat called blubber. Short, stiff hair on the paws helps keep the animal's feet warm as they pad over ice and snow.

When you go outside on a cold day, you need to bundle up in extra clothing to stay warm. The polar bear is already dressed to survive in its wintry world. ■

JOAN MARCUS

Why is it dangerous to touch dry ice?

E.R. DEGGINGER

White fog surrounds actors on a stage (left). Hidden from the audience, technicians are creating the mysterious-looking fog. They do it by feeding chunks of dry ice—solid carbon dioxide—into hot water.

The fog itself is not dangerous, so the actors do not need to take any special precautions. But the temperature of the dry ice that produces the fog is lower than -100°F. The dry ice is so cold that it can cause frostbite. The technicians who handle it must wear heavy gloves for protection.

Ordinary ice melts into a liquid as it warms up. Dry ice—as in the photograph above—can change directly from a solid to a gas. Shippers often pack food in dry ice to keep it cold.■

How can water exist as a solid, a liquid, and a gas in the same environment?

When you hear the word "snowsuit," you don't normally think of a swimsuit. Yet swimsuits are all these boys wear as they play in a pile of snow. When they feel chilled, they dash through mist and jump into the heated water of a nearby swimming pool. The boys are enjoying different forms of water all at once.

Liquid water can change into water vapor, a gas, in a process known as evaporation. Solid water, or ice, can change directly into water vapor in a process known as sublimation. Both processes can occur at various temperatures, but they speed up as the temperature rises.

Wherever liquid water or ice exists, water vapor also exists. It is invisible. In this pool, the water's warm temperature has speeded up the process of evaporation. The water vapor has condensed, or turned into a liquid, which you see as mist in the air.

Liquid water freezes, and ice melts, at 32°F. Nevertheless, ice and snow—a form of ice—can exist at temperatures above 32°F. A lot of energy is needed to melt even a small amount of ice or snow. The warm water in the swimming pool does contain a lot of energy, but it releases that energy very slowly. Air conducts energy poorly. As a result, not much energy from the hot water reaches the nearby snow. The snow and warm water can continue to exist side by side until the air temperature rises enough to melt the snow.■

WINFIELD PARKS

THE EYES HAVE IT

How do eyes see?

Learning about vision can be an eye-opening experience.

Your eyes work like lookouts, gathering information from rays of light. They pass the information to the brain, which translates it into shapes and colors, sizes and distances.

Like any complicated machine, the eye has many different parts. One of the parts is the sclera (SKLEHR-uh), the white of the eye. The colored part of the eye is called the iris. The cornea (KOR-nee-uh) covers the iris like a transparent window. The pupil, which appears black, is a hole in the center of the iris. The pupil lets light through the lens, which lies behind the iris. The retina (RET-nuh) lines the inside of the eye. Nerve cells called rods and cones in the retina help make vision possible. The optic nerve connects the retina to the brain.

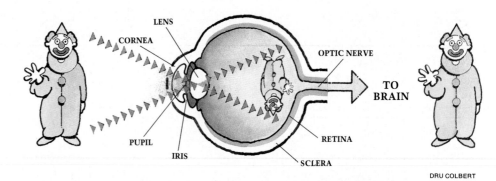

DRU COLBERT

Imagine you are looking at a clown (above). Light bounces off the clown and passes through the cornea. The cornea bends that light toward the center of the eye, and the light passes through the pupil. Muscles in the iris adjust the size of the pupil to allow just the right amount of light to reach the lens. In dim light, for example, the pupil expands to allow as much light as possible to enter. The lens focuses light rays onto the retina and projects an image of the clown that is upside down and reversed left to right.

When light rays strike the retina, rods and cones change them into signals for the brain. The optic nerve transmits the signals, and the brain translates them into shapes and colors. The brain turns the clown image right side up and right to left.

How long does this complicated process take? No longer than it takes to blink an eye!■

Why do eyes sometimes look red in flash pictures?

This figure seems to have stepped out of a fantasy or a fairy tale. In fact, she's a normal human being. Photography has made her look for a moment like a strange creature with red eyes.

Blood vessels in the retina at the back of the eye are red. When a photographer aims a camera with a flash attachment directly at a person's face, the flash shines through the pupil and lights up the retina. The camera records the red color on film. In this picture, the type of film the photographer used also makes the model's skin and the whites of her eyes appear bluish.

If you use a camera, you probably know that "red eye" is common in flash photography. It results when a camera's flash attachment is close to the camera's lens, and both flash attachment and lens are on the same level as the eyes of the person in the photograph. Here are a few ways to avoid red eye in pictures you make. If possible, move the flash attachment away from the camera lens. Tilt your flash upward. Or ask the people in your picture to look away from the camera slightly before they smile and say, "Cheese!"■

DORLING KINDERSLEY LTD., LONDON/TREVOR MELTON

72

Why do some animals' eyes seem to glow in the dark?

The eyes of a wildcat appear to glow like twin lamps. The "glow" is actually a reflection of a light directed toward the animal's face.

Some animals have in their eyes a layer of cells called the *tapetum* (tuh-PEA-tum). Like many tiny mirrors, these cells reflect light. Some scientists think the reflected light may improve the animals' night vision. The light has been described as shining like highly polished metal. It also shines in different colors—often in shades of blue, green, gold, or silver. That's because coloring materials in the tapetum may vary from one animal to another.

Scientists refer to the shimmering sight as eyeshine. Besides cats, animals with eyeshine include dogs, raccoons, bullfrogs, horses, foxes, deer, and even some insects.■

Why are some people color blind?

Take a good look at these circles filled with colorful dots. Whether you have normal color vision or not, you should be able to see the number **12** in the top circle. If you are color blind, however, you may not be able to make out the number **5** in the bottom circle.

Your eyes contain two kinds of light-sensitive cells: rods and cones. Rods enable you to see in shades of gray. Cones enable you to see in colors. In some people, the cones are defective. In other people, information gathered by the cones is not passed properly from the eyes to the brain. Both kinds of people are color blind.

People can inherit color blindness from their parents. In the most common form of color blindness, people have difficulty telling the difference between the colors red and green. A girl inherits this condition from both her mother and father. A boy can inherit this condition from his mother alone. That's why more boys than girls are color blind.

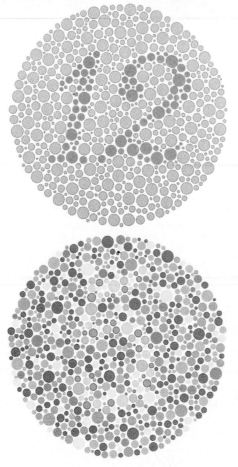

GRAHAM-FIELD, INC.

In rare cases, people are totally color blind. The cones in their eyes do not function at all.

Do you have questions about your own color vision? If so, you might wish to check with an eye doctor.■

Why do owls see well in the dark?

Late at night, a screech owl heads toward its nest. In its beak it carries a beetle it has caught for a midnight meal. Sharp night sight helps the owl capture prey in darkness.

An owl has big eyes for an animal of its size. The eyes of a great horned owl, for example, are as big as the eyes of a fully grown person, although the owl is much smaller than a person. In near darkness, an owl's big eyes allow available light to reach the retina.

The more rods an animal has in its eyes, the better its night vision. An owl's eyes have many more rods than cones. In fact, the difference between the number of rods and cones in the eyes of an owl is greater than it is in the eyes of many other animals.

Sharp hearing reinforces sharp eyesight when an owl hunts at night. An owl picks up the sound of its prey on the move — even sounds as soft as the scampering of a mouse.■

JOHN F. O'CONNOR, M.D.: PHOTO/NATS

PAINFUL NEWS

Why does my side sometimes hurt when I run?

This runner has run into a problem: a pain in her side. The pain is a form of muscle cramp called a "stitch."

Doctors aren't sure what causes stitches. Running puts pressure on a breathing muscle called the diaphragm (DY-uh-fram). The pressure may cause a stitch. Gas or cramps in the intestines may be another cause of the pain. If so, avoid eating or drinking anything right before you run and you may be able to avoid stitches.

Anyone—even a highly trained athlete—can suffer from stitches. If you feel the pain of a stitch, relax. It won't last long. Slow down or stop running for a few minutes, and the feeling should fade.■

Why do I get black-and-blue marks?

Crrrrunch. Soccer players collide on a field (right). They may have quite a few black-and-blue marks, or bruises, by the time they finish their game.

Bruises can form on your body anyplace where you are hit hard, but not hard enough to break the skin. The injury causes small blood vessels called capillaries to burst and bleed *under* the skin. The blood darkens. Through your skin, you may see the blood as a black-and-blue mark.

A bruise on certain parts of your body can turn into a bump. In fleshy areas like your thigh, you have more tissue between bone and skin to absorb swelling. But on your elbow, head, or shin, bones lie close to the skin. There is less tissue to cushion a blow and less space for blood to collect. If you are hit in one of these places, a bump may rise to make room for blood trapped under the skin.

Fortunately, ordinary bumps and bruises heal and disappear without special care.■

What causes blisters?

One of these soccer players may end up with more than bumps and bruises. If his shoes don't fit comfortably, they may rub the skin on his feet. The rubbing irritates blood vessels just beneath the skin as well as skin cells. Fluid leaks from those blood vessels and skin cells. It builds up in a blister between the top layer of skin and the underlying layer. Eventually the top layer dies, and new skin replaces it.

Blisters can also form on hands irritated by rubbing, or anywhere on the body where skin has been burned.■

© BOB DAEMMRICH

BOB DAEMMRICH/UNIPHOTO

EXPRESSIONISTS

Why do people laugh?

Laughter, some people say, is the best medicine. If that's true, then these young soccer players sharing a friendly laugh should be very healthy.

Laughter releases emotions through physical movement. When you laugh, you breathe in deeply, then exhale in quick bursts. The air that you push out of your lungs passes by the larynx (LAR-inx), or voice box, in your throat. Two bands called vocal cords are located in the larynx. Muscles draw the vocal cords together, and air from the lungs causes them to vibrate. Out come the sounds of laughter.

Most people, of course, don't laugh to exercise their vocal cords. You might laugh because you feel happy, or playful, or because you just heard a funny joke. You might laugh at yourself for making a silly mistake, or at a friend who is clowning around. Laughter can be contagious: You might laugh because you see someone else laughing. You might even laugh because you feel nervous or embarrassed. One writer has listed 80 different reasons why people laugh!■

Why do babies often cry?

ENRICO FERORELLI/DOT

The tears of a baby are no laughing matter. Babies have needs just as older people do. But a baby can't describe those needs. A baby can't say, "I'm hungry. Feed me, please. Then change my diaper and give me a big hug."

The only way a newborn baby can communicate is by crying. The tears of all babies carry a message. If the person taking care of a healthy baby can correctly guess what the message means and can satisfy the baby's needs, the tears will probably stop.

If you have a new baby brother or sister, don't worry. You'll soon hear gurgling and cooing as well as shrieking and whimpering. By the first birthday, many babies can say a word or two. By the time they are five years old, most children speak well, and words usually take the place of tears.■

TO YOUR HEALTH!

How does sugar cause cavities?

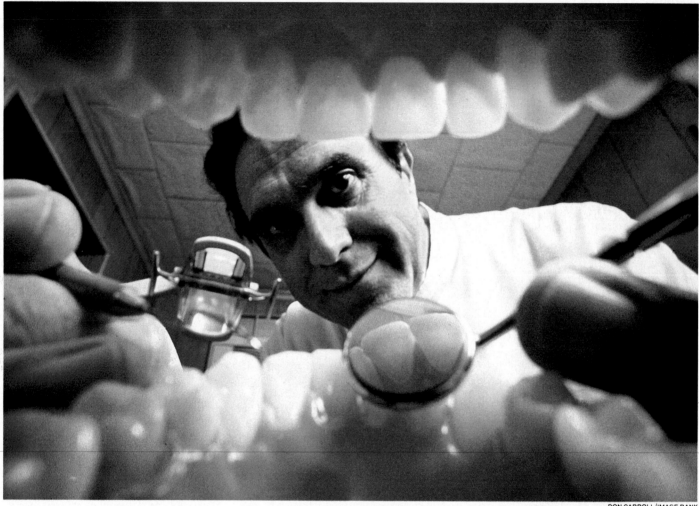

DON CARROLL/IMAGE BANK

"Open wide." The dentist checks your teeth for cavities, or small holes. If you have any, the cause may be sugar.

The trouble starts, most dentists say, when bacteria in your mouth feed on tiny food particles stuck between your teeth, producing an acid. The bacterial reaction with sugar creates more acid than the reaction with most other kinds of food.

The acid can slowly eat a hole through the layer of hard enamel that coats all of your teeth. At first, the cavity in a tooth may be small. But if left untreated, it can grow bigger. Eventually, the inside of your tooth may begin to decay too, and you may find yourself with a painful toothache.

You can help prevent cavities by eating fewer sugary foods and by cleaning your teeth regularly. Brush them at least twice a day and clean between them once a day with dental floss. When you brush and floss, you remove many food bits before bacteria can produce acid. That should give you something to smile about!■

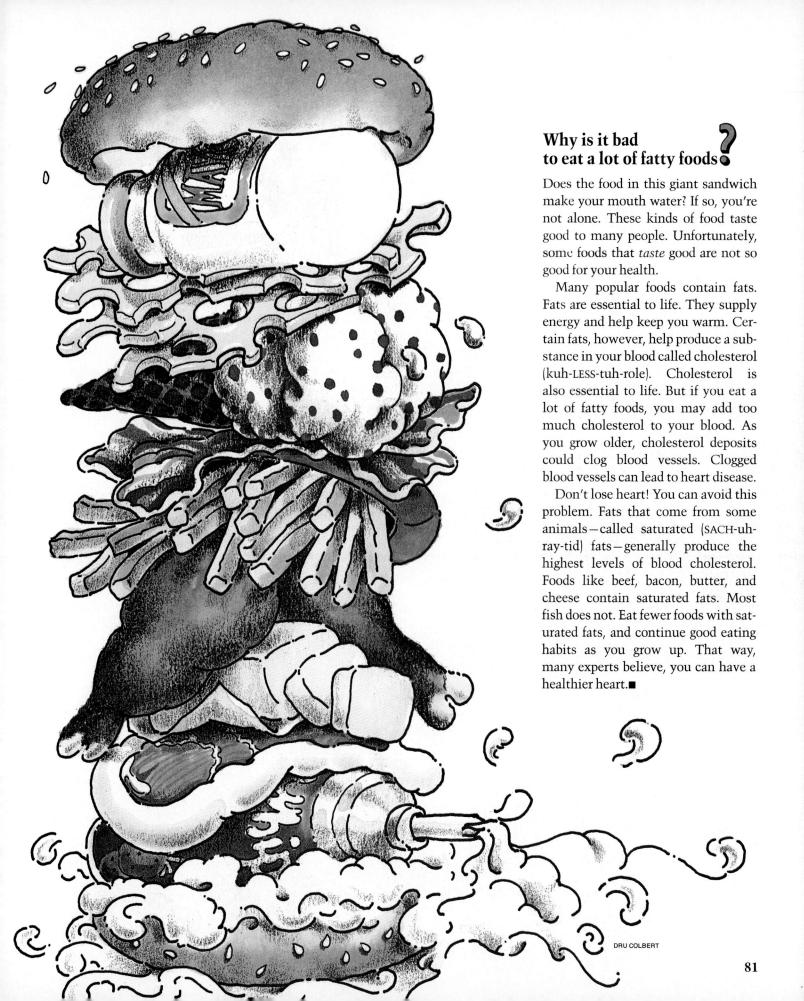

Why is it bad to eat a lot of fatty foods?

Does the food in this giant sandwich make your mouth water? If so, you're not alone. These kinds of food taste good to many people. Unfortunately, some foods that *taste* good are not so good for your health.

Many popular foods contain fats. Fats are essential to life. They supply energy and help keep you warm. Certain fats, however, help produce a substance in your blood called cholesterol (kuh-LESS-tuh-role). Cholesterol is also essential to life. But if you eat a lot of fatty foods, you may add too much cholesterol to your blood. As you grow older, cholesterol deposits could clog blood vessels. Clogged blood vessels can lead to heart disease.

Don't lose heart! You can avoid this problem. Fats that come from some animals—called saturated (SACH-uh-ray-tid) fats—generally produce the highest levels of blood cholesterol. Foods like beef, bacon, butter, and cheese contain saturated fats. Most fish does not. Eat fewer foods with saturated fats, and continue good eating habits as you grow up. That way, many experts believe, you can have a healthier heart.■

DRU COLBERT

THE HOLE TRUTH

How can a sponge hold liquid if it's full of holes?

SUSAN M. KLEMENS (BOTH)

Oddly enough, a sponge with holes can hold liquid better than it could if it had no holes. The holes, or pores, expose more of the sponge's absorbent surface area to the liquid. Manufacturers put holes in sponges on purpose to improve their product.

For centuries, people have been cleaning up with natural sponges—skeletons of simple animals called sponges that live in the ocean. Today, people usually rely on synthetic sponges. Manufacturers make most of these sponges from an absorbent fiber called cellulose (SELL-yoo-lohs) which they obtain mainly from trees. These youngsters use cellulose sponges to wash the family car. Soapsuds fly (below) as they tackle the job.

To make sponges, manufacturers mix up a cellulose solution. They add chemicals to the mixture to change its composition. They add natural fibers to strengthen it, and salt crystals to form the pores. The size of the crystals determines the size of the pores. Dye is added for coloring. Then workers bake the mixture in an oven. The cooked cellulose mixture is washed, softened, and sliced into sponges. Some cellulose sponges can hold 20 times their weight in fluids!■

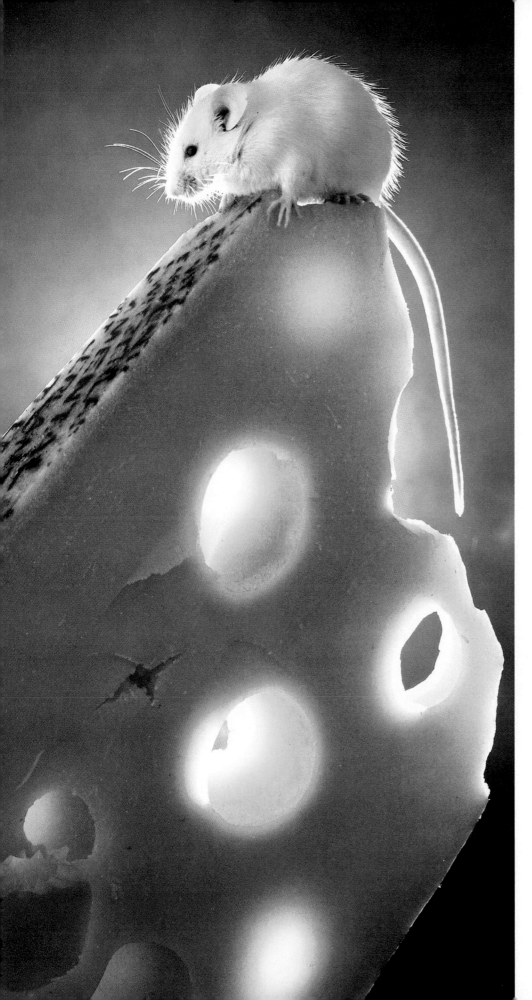

Why do holes form in Swiss cheese?

What has eyes but cannot see? Swiss cheese. Cheesemakers refer to the holes in Swiss cheese and other kinds of cheeses as "eyes." They learn to predict the quality of a cheese by studying the eyes. This lucky mouse perches on top of a chunk of Swiss cheese in a photographer's studio.

All natural cheeses come from milk. Workers treat the milk to form a thick substance called curd, and they separate from the curd a liquid called whey. During the first part of this process, cheesemakers who are preparing Swiss cheese add different kinds of bacteria. One kind helps produce the familiar taste of Swiss cheese. It also produces bubbles of carbon dioxide gas. The carbon dioxide forms holes in the cheese.

Holes usually begin to appear when Swiss cheese is about three weeks old. In the United States, cheesemakers allow Swiss cheese to ripen for at least two months before they sell it to consumers. In Switzerland, cheesemakers often let the cheese ripen much longer. The longer the ripening period, the sharper the flavor.■

One Horn or Two?

Why did some people believe unicorns were real?

For centuries, people believed in the existence of a swift, white, horse-like animal with a single horn growing from its head. Everyone from kings to peasants heard stories about the beast, named "unicorn" from the Latin words meaning "one horn." According to some stories, a unicorn that came upon a young girl in the forest would kneel by her side. Then hunters could capture the resting animal.

Many people thought a unicorn's horn had magical powers to cure diseases, detect poison, and purify water. Traders sold "unicorn horns" at high prices. Artists wove images of the

84

unicorn into tapestries (right, bottom).

Today experts agree that real unicorns did not exist. Researchers think the horns that traders sold did come from real animals, however—perhaps from rhinoceroses or from narwhals, small whales with a long ivory tusk.

Stories of the unicorn may have been based on sightings of rhinoceroses or other animals. Some travelers of long ago may have seen sheep or antelopes with only one horn. Maybe the other horn had broken off, or maybe the animals were freaks of nature called "sports," born with one horn instead of the usual two. Perhaps animals with two horns appeared to travelers to have only one. If you look at the antelopes on these pages, you can understand how easy it would be for this to happen. Books, art, and travelers' tales helped convince people that the stories of unicorn sightings were true.■

PET THEORIES

Why does a dog wag its tail?

When it comes to dogs, the tails tell quite a tale. The way a dog moves its tail sends a message: "Glad to see you," perhaps, or "Move over, Rover." A dog uses the same kind of tail movement to send a message to a person or to another dog.

A frightened dog carries its tail between its legs. A wagging tail extended straight out or upright is a likely sign that a dog is feeling friendly.

If you approach a strange dog, pay attention to its tail action. If the dog holds its tail still, it is not sure how to react to you. If the dog then starts to wag its tail in a low position, it probably won't hurt you. But if the animal arches its tail, or holds it stiff and high without wagging, watch out! Such tail signals could mean trouble. A dog with its tail in one of these positions does not want to make friends. Back away slowly before Fido turns fierce. ■

DRU COLBERT

LON E. LAUBER/WEST STOCK, INC.

Why is a dog's nose cold?

You and your pet pooch are playing catch on a summer day. The exercise makes the dog hot, but its nose stays cool. That cold nose is a sign that the animal's temperature-control system is working.

Sweat glands in a dog's nose send moisture to the surface of the skin there. The glands produce sweat when the animal is active. As the sweat evaporates from the nose, it carries heat away from the body and helps cool the dog. If you touch the dog's nose, it feels cold.

Since the nose of a healthy dog often feels cold and wet, many people assume that a dog with a warm, dry nose must be sick. Although this is sometimes the case, it isn't always so. A dog that has been sleeping for a while will usually wake up with a warm, dry nose. The dog isn't ill, just resting. Its nose will probably feel cold and wet again when the dog becomes active. ■

Why do cats have whiskers?

If you've ever dressed up in a cat costume for Halloween, you've probably decorated your face with whiskers. For a real cat, whiskers provide information, not decoration.

Whiskers, or vibrissae (vy-BRIHS-ee), grow on a cat's upper lip, chin, and cheeks. More whiskers grow above the animal's eyes and behind its front legs. Sensitive nerves lie at the root of every whisker. When something touches a whisker, the nerves send signals to the cat's brain. The signals give the cat information about what's going on around it. They might, for example, help the cat judge if it is too big to squeeze through a narrow opening.

Whiskers also detect air currents bouncing off solid objects, such as furniture inside a house or bushes outside. During the night, this helps a cat sense obstacles that it might not see well in darkness, so it can move around more safely.■

SPIN CYCLE

Do skaters and dancers get dizzy when they spin?

They whirl. They twirl. Expert ice skaters and ballet dancers can spin around and around so fast they almost make you dizzy. But *they* may not feel dizzy. Constant practice helps them overcome dizziness.

Ballet dancers also learn a method called *spotting.* They choose a spot to fix their eyes on. At the start of a turn, they stare at the spot. Then they whip the head around, and quickly return their gaze to the same spot. The head completes the turn before the rest of the body does.

In a multiple-image photograph, ballet dancer Vanessa Tiegs (right) turns with one leg raised. Spotting helps her turn again and again without getting dizzy.

Ice skaters aren't so lucky. They spin much faster than ballet dancers do. If they tried to stare at one spot while turning, they would risk serious neck injury. When an ice skater spins, the head turns with the rest of the body. Beginning skaters often do get dizzy when learning how to spin. Practice helps. Experienced skaters concentrate on keeping their balance. As they begin to slow down and come out of a spin, they may focus on one object to reduce dizziness. Eventually they adjust to the rapid twirling and the dizziness it brings.

Below, left to right, are figure skaters Junko Yaginuma, of Japan, and Paul Wylie and Debi Thomas, of the United States. The three super spinners competed at the 1988 Winter Olympics, which were held in Calgary, Alberta, in Canada. ∎

ALLSPORT PHOTOGRAPHY/RUSSELL CHEYNE

BILL EPPRIDGE/SPORTS ILLUSTRATED

BILL EPPRIDGE/SPORTS ILLUSTRATED

FOR THE BIRDS

Why don't ducks sink?

Under their mother's watchful eye, ducklings sail forth to explore the water. A duck can swim and float with ease even though its body is heavier than water.

Tiny projections called barbules (BAR-byoolz) on a duck's feathers hook together and trap air between the feathers. Air sacs connected to the lungs hold more air inside the body. Air between the feathers and in the air sacs helps keep a duck on top of the water. A duck can partly deflate the air sacs when it dives to look for food.

A duck spends a lot of time preening its feathers—arranging them with its beak and coating them with oil from a gland above its tail. Because preening waterproofs feathers, many scientists think that it helps a duck float.■

SCOTT NIELSEN/BRUCE COLEMAN INC.

SHARON DAVIS

Why do many birds "get up" early in the morning?

When it's time for a wake-up call, the early birds can make quite a racket.

Early morning light awakens many birds. They tend to wake earlier on sunny mornings than they do on cloudy or rainy days. Once awake, these birds stay active throughout the early morning. They sing, fly around searching for food, and, during breeding season, they mate. They often rest when the temperature rises at midday, then become busy again from late afternoon until sunset.

Not all birds are early birds. Some owls hunt at night and sleep the day away. Not all early birds are birds, either. Other kinds of animals also wake up with the rising sun.■

Why do birds grow new feathers ❓

A fuzzy "wig" of brown down feathers grows from the head of an albatross chick. Newer white feathers have already grown in and taken the place of down on the bird's chest. Soon they will replace down on the head as well.

Most birds molt—lose old feathers and grow new ones—at least once a year. A thick coat of feathers may grow during winter and provide warmth. These feathers may be replaced by more colorful ones that attract mates during breeding season.

In most birds, flight feathers drop out gradually. Ducks, however, lose their flight feathers all at once. Molting ducks cannot fly. They find food—and safety—in the water.■

GOING OUT WITH A BANG

Why do I usually see fireworks before I hear their noise?

Fireworks explode over New York's East River on July 4, 1986 (right). The display celebrates the restoration of the Statue of Liberty.

When you watch a fireworks display, you may notice that light from the explosions reaches you almost instantly. Sound from the explosions lags slightly behind the light. That's because light travels almost a million times faster than sound through air.

There may be an additional explanation for seeing the blast before hearing the bang. Inside some fireworks are pellets called *stars* and a large firecracker known as a *salute*. Chemicals that make up the stars burn with bright colors. Salutes contain a chemical mixture that explodes loudly. Manufacturers often design fireworks to burst with colorful stars, then explode with a noisy salute.■

Why do fireworks explode in different colors?

Early fireworks gave off only fiery sparks and flames. Modern fireworks explode in many colors—all created by chemicals that produce different shades when heated in a flame. Strontium nitrate or strontium carbonate produces a red flame. Sodium salts burn yellow. Barium nitrate or chlorate glows green. Some copper compounds produce certain shades of blue—but fireworks designers are still searching for just the right mixture of chemicals that will burst into a flame of deep electric blue.■

ROBERT D. RUBIC/DPI

INDEX

Bold type refers to illustrations; regular type refers to text.

Library of Congress ℂℙ Data
Why on Earth?
(Books for world explorers)
Includes index.
Summary: Questions and answers present information on science topics, including human physiology, animal behavior, earth science, and natural science.
1. Science—Miscellanea—Juvenile literature. [1. Science—Miscellanea. 2. Questions and answers] I. National Geographic Society (U.S.) II. Series.
Q163.W494 1988   500   88-25486
ISBN   0-87044-701-7 (regular edition)
ISBN   0-87044-706-8 (library edition)

CONSULTANTS

Glenn O. Blough, LL.D., Emeritus Professor of Education, University of Maryland, *Educational Consultant*

Colleen Conley, M.D., Children's Hospital National Medical Center, George Washington University, *Consulting Psychiatrist*

Joan Winchester Myers, M.Ed., Alexandria City Schools, *Reading Consultant*

The Special Publications and School Services Division is also grateful to the individuals and institutions named or quoted in the text and to those cited here for their generous assistance:

Chester W. Anderson, D.V.M., Peachtree Veterinary Clinic; John R. Apel, Ph.D., The Johns Hopkins University; Robert T. Bakker, Ph.D., University of Colorado Museum; Joseph F. Barbera, O-Cel-O Sponge; Bonnie Beaver, D.V.M., Texas A & M University; Susan Bell, Bill Lewis, Nancy Schaadt, John F. Kennedy Center for the Performing Arts; Richard E. Berg, Ph.D., University of Maryland; Eddie N. Bernard, Ph.D., National Oceanic and Atmospheric Administration; Marilyn Berzin, M.D., Washington, D.C.; Denise Breitburg, Ph.D., Benedict Estuarine Research Laboratory; Michael Brett-Surman, Elvira Clain-Stefanelli, Paul W. Pohwat, Lynn Vosloh, Smithsonian Institution; B.G. Bricks, Ph.D., W.J. Schafer Associates; Bruce Britts, University of Sydney, Sydney, Australia.

Tristram Potter Coffin, Ph.D., University of Pennsylvania; John Conkling, Ph.D., American Pyrotechnics Association; Raymon G. Davie, Cooperstown, New York; Joanne Economon, M.D., Washington, D.C.; Albert W. Erickson, Ph.D., University of Washington; Bernard Fitzmorris, D.D.S., Washington, D.C.; Jacqueline Geschickter, Virginia Ballet Company and School; Paul Gipe, American Wind Energy Association; Carrie Hulse, Greater Juneau Chamber of Commerce; Seth Koch, D.V.M., Animal Eye Clinic; Walter Krahenbuhl, Washington Cheese Inc.; Ian Lancaster, Janine W. MacKenna, Museum of Holography; Gerald Loughlin, M.D., The Johns Hopkins Hospital; Stephen P. Maran, Ph.D., NASA Goddard Space Flight Center; Christopher J. Murphy, D.V.M., University of California.

Gregory S. Paul, Baltimore, Maryland; Annie Paulson, National Wildflower Research Center; Andrew Pogan, Montgomery County Schools, Maryland; Vernon G. Pursel, Ph.D., Steve Sheppard, Ph.D., U.S. Department of Agriculture; Richard B. Reff, M.D., Children's Hospital National Medical Center; Marianne Schuelein, M.D., Georgetown University Medical Center; Peter B. Stifel, Ph.D., University of Maryland; Fiona Sunquist, Melrose, Florida; Bruce W. Turner, Pacific Tsunami Warning Center; Bradford Washburn, Boston Museum of Science; George E. Watson, Ph.D., Saint Albans School; Knox Williams, Colorado Avalanche Information Center; Owen W. Williams, Annandale, Virginia; David Zeiler, D.V.M., Cabin John, Maryland.

Composition for WHY ON EARTH? by the Typographic section of National Geographic Production Services, Pre-Press Division. Type mechanicals by Carrie A. Edwards and Marvin J. Fryer. Printed and bound by Holladay-Tyler Printing Corp., Glenn Dale, Md. Film preparation by Catharine Cooke Studio, Inc., New York, N.Y. Color separations by Lanman-Progressive Co., Washington, D.C., and NEC, Inc., Nashville, Tenn. Cover printed by Federated Lithographers-Printers, Inc., Providence, R.I. Teacher's Guide printed by McCollum Press, Inc., Rockville, Md.

WHY ON EARTH?

PUBLISHED BY
THE NATIONAL GEOGRAPHIC SOCIETY
WASHINGTON, D.C.

Gilbert M. Grosvenor, *President and Chairman of the Board*
Melvin M. Payne, Thomas W. McKnew, *Chairmen Emeritus*
Owen R. Anderson, *Executive Vice President*
Robert L. Breeden, *Senior Vice President, Publications and Educational Media*

PREPARED BY THE SPECIAL PUBLICATIONS AND SCHOOL SERVICES DIVISION

Donald J. Crump, *Director*
Philip B. Silcott, *Associate Director*
Bonnie S. Lawrence, *Assistant Director*

BOOKS FOR WORLD EXPLORERS

Pat Robbins, *Editor*
Ralph Gray, *Editor Emeritus*
Ursula Perrin Vosseler, *Art Director*
Margaret McKelway, *Associate Editor*
Larry Nighswander, *Illustrations Editor*

STAFF FOR *WHY ON EARTH?*

Susan Mondshein Tejada, *Managing Editor*
Susan M. Klemens, *Picture Editor*
Louise Ponsford, *Art Director*
Barbara L. Bricks, Valerie A. May, *Researchers*
Sharon L. Barry, Jan Leslie Cook, Robin Darcey Dennis, Ann DiFiore, Betty Mussell Lundy, Susan McGrath, Toni Rizzo, Elizabeth H. Summers, Suzanne Venino, Karen Romano Young, *Writers*
Roger B. Hirschland, *Consulting Editor*
M. Barbara Brownell, Jane R. McGoldrick, *Contributing Editors*
John G. Agnone, *Consulting Picture Editor*
Mary Lee Elden, *Contributing Researcher*
Kathryn N. Adams, Sandra F. Lotterman, *Editorial Assistants*
Janet A. Dustin, Jennie H. Proctor, *Illustrations Assistants*
Michele L. du Vair, *Illustrations Intern*

ENGRAVING, PRINTING, AND PRODUCT MANUFACTURE: George V. White, *Director;* Vincent P. Ryan, *Manager;* David V. Showers, *Production Manager;* Lewis R. Bassford, *Production Project Manager;* Kathie Cirucci, Timothy H. Ewing, *Senior Production Assistants;* Kevin Heubusch, *Production Assistant;* Carol R. Curtis, *Senior Production Staff Assistant*

STAFF ASSISTANTS: Aimée L. Clause, Catherine G. Cruz, Marisa Farabelli, Mary Elizabeth House, Rebecca A. Hutton, Karen Katz, Lisa A. LaFuria, Eliza C. Morton, Dru Stancampiano, Nancy J. White

MARKET RESEARCH: Joseph S. Fowler, Carrla L. Holmes, Marla Lewis, Joseph Roccanova, Donna R. Schoeller, Marsha Sussman, Judy T. Guerrieri

INDEX: Lucinda L. Smith